Collins

Year 7, Practice Book 2

NEW MATHS FRAMEWORKING

Matches the revised KS3 Framework

Kevin Evans, Keith Gordon, Trevor Senior, Brian Speed

William Collins' dream of knowledge for all began with the publication of his first book in 1819. A self-educated mill worker, he not only enriched millions of lives, but also founded a flourishing publishing house. Today, staying true to this spirit, Collins books are packed with inspiration, innovation and practical expertise. They place you at the centre of a world of possibility and give you exactly what you need to explore it.

Collins. Do more.

Published by Collins
An imprint of HarperCollins*Publishers*
77–85 Fulham Palace Road
Hammersmith
London
W6 8JB

Browse the complete Collins catalogue at
www.collinseducation.com

© HarperCollins*Publishers* Limited 2008

10 9 8 7

ISBN 978-0-00-726792-7

Keith Gordon, Kevin Evans, Brian Speed and Trevor Senior assert their moral rights to be identified as the authors of this work.

All rights reserved. No part of this publication may be reproduced, stored in a retrieval system, or transmitted in any form or by any means, electronic, mechanical, photocopying, recording or otherwise, without the prior written permission of the Publisher or a licence permitting restricted copying in the United Kingdom issued by the Copyright Licensing Agency Ltd., 90 Tottenham Court Road, London W1T 4LP.

British Library Cataloguing in Publication Data
A Catalogue record for this publication is available from the British Library.
Commissioned by Melanie Hoffman and Katie Sergeant
Project management by Priya Govindan
Covers management by Laura Deacon
Edited by Brian Ashbury
Proofread by Amanda Dickson
Design and typesetting by Newgen Imaging
Design concept by Jordan Publishing Design
Covers by Oculus Design and Communications
Illustrations by Derek Lee and Newgen Imaging
Printed and bound by Printing Express, Hong Kong
Production by Simon Moore

Every effort has been made to trace copyright holders and to obtain their permission for the use of copyright material. The authors and publishers will gladly receive any information enabling them to rectify any error or omission in subsequent editions.

MIX
Paper from
responsible sources
FSC® C007454

FSC™ is a non-profit international organisation established to promote the responsible management of the world's forests. Products carrying the FSC label are independently certified to assure consumers that they come from forests that are managed to meet the social, economic and ecological needs of present and future generations, and other controlled sources.

Find out more about HarperCollins and the environment at
www.harpercollins.co.uk/green

Introduction

Welcome to *New Maths Frameworking*!

New Maths Frameworking Year 7 Practice Book 2 has hundreds of levelled questions to help you practise Maths at Levels 4-5. The questions correspond to topics covered in Year 7 Pupil Book 2 giving you lots of extra practice.

These are the key features:

- **Colour-coded National Curriculum levels** for all the questions show you what level you are working at so you can easily track your progress and see how to get to the next level.

- **Functional Maths** is all about how people use Maths in everyday life. Look out for the Functional Maths icon (FM) which shows you when you are practising your Functional Maths skills.

Contents

CHAPTER 1 Algebra **1** — 1

CHAPTER 2 Number **1** — 3

CHAPTER 3 Geometry and Measures **1** — 7

CHAPTER 4 Number **2** — 10

CHAPTER 5 Statistics **1** — 14

CHAPTER 6 Algebra **2** — 19

CHAPTER 7 Geometry and Measures **2** — 24

CHAPTER 8 Statistics **2** — 26

CHAPTER 9 Number and Measures **3** — 29

CHAPTER 10	Algebra 3	33
CHAPTER 11	Geometry and Measures 3	38
CHAPTER 12	Number 4	40
CHAPTER 13	Algebra 4	43
CHAPTER 14	Geometry and Measures 4	45
CHAPTER 15	Statistics 3	50
CHAPTER 16	Number 5	53
CHAPTER 17	Algebra 5	56
CHAPTER 18	Geometry and Measures 5	60

CHAPTER 1 Algebra 1

Practice 1A Sequences and rules

1 Use each of the following term-to-term rules with 1st term 2. Create the sequences with 5 terms each.

a add 3
b add 6
c treble
d multiply by 5
e add 100
f multiply by 10

2 Write the next two terms in each sequence. Describe the term-to-term rule you have used.

a 1, 3, 5, 7, …
b 20, 30, 40, 50, …
c 5, 13, 21, 29, …
d 5, 10, 15, 20, …
e 6, 13, 20, 27, …
f 10, 110, 210, 310, …

3 Fill in the gap to make at least one sequence, fully describing the term-to-term rule you have used.

a 3, …, 7
b 1, …, 9
c 10, …, 50
d 2, …, 14
e 6, …, 16
f 20, …, 100

4
a Add any two sequences from question 2, term-by-term.
b Write down the 1st term and term-to-term rule.
c You could answer part **b** without adding the sequences. How?

Practice 1B Finding missing terms

1 In each of the following sequences, find the 5th and 50th terms.

a 1, 5, 9, 13, …
b 3, 5, 7, 9, …
c 4, 12, 20, 28, …
d 5, 15, 25, 35, …
e 2, 8, 14, 20, …
f 10, 30, 50, 70, …
g 2, 5, 8, 11, …
h 0, 5, 10, 15, …
i 4, 11, 18, 25, …

2 In each of the following sequences, find the missing terms and the 30th term.

Term	1st	2nd	3rd	4th	5th	6th	7th	8th	…	30th
Sequence A	__	__	__	13	16	19	22	__	…	__
Sequence B	__	9	16	__	30	37	__	__	…	__
Sequence C	__	__	25	__	45	__	65	__	…	__
Sequence D	__	11	__	19	__	27	__	__	…	__

Practice: 1C Functions and mappings

1) Use the function to complete the output for each of the following function machines.

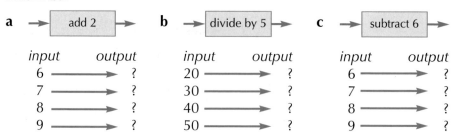

a) add 2
- input 6 → ?
- input 7 → ?
- input 8 → ?
- input 9 → ?

b) divide by 5
- input 20 → ?
- input 30 → ?
- input 40 → ?
- input 50 → ?

c) subtract 6
- input 6 → ?
- input 7 → ?
- input 8 → ?
- input 9 → ?

2) Express these simple functions in words.

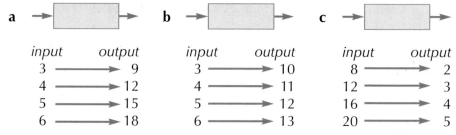

a)
- input 3 → output 9
- input 4 → output 12
- input 5 → output 15
- input 6 → output 18

b)
- input 3 → output 10
- input 4 → output 11
- input 5 → output 12
- input 6 → output 13

c)
- input 8 → output 2
- input 12 → output 3
- input 16 → output 4
- input 20 → output 5

3) Draw diagrams to illustrate the following functions. Start with any numbers you like for the input, but remember, the larger the numbers the more difficult the problem.

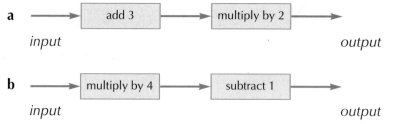

a) input → add 3 → multiply by 2 → output

b) input → multiply by 4 → subtract 1 → output

4) Write down the inverse function of:

a) subtract 8   b) multiply by 4

5) Each of the following functions are made up from two operations as above. Find the combined functions.

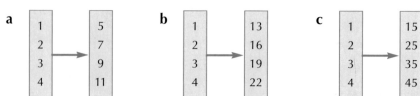

a) 1→5, 2→7, 3→9, 4→11

b) 1→13, 2→16, 3→19, 4→22

c) 1→15, 2→25, 3→35, 4→45

Practice 1D Using letter symbols to represent functions

1. Write down what the expression $x + 10$ is equal to when:
   i $x = 3$ ii $x = 10$ iii $x = 18$ iv $x = 50$ v $x = 120$

2. Write down what the expression $4n$ is equal to when:
   i $n = 4$ ii $n = 1$ iii $n = 7$ iv $n = 12$ v $n = 0$

3. Write the following rules in symbolic form, for example, $x \to x + 4$.
   a subtract 5   b treble   c add 9   d divide by 3

4. Draw mapping diagrams to illustrate the following functions:
   a $x \to x + 25$ b $x \to 3x$ c $x \to 3x - 1$ d $x \to 4x + 4$

5. Express the following functions in symbols as in question 3.

   a
8	$\to$	12
9	$\to$	13
10	$\to$	14
11	$\to$	15

   b
2	$\to$	12
3	$\to$	18
4	$\to$	24
5	$\to$	30

   c
1	$\to$	5
2	$\to$	7
3	$\to$	9
4	$\to$	11

   d
1	$\to$	1
2	$\to$	4
3	$\to$	7
4	$\to$	10

Practice 1E The general term [nth term]

1. Find **i** the first three terms and **ii** the 100th term, of sequences whose nth term is given by:
   a $3n - 1$
   b $5n + 2$
   c $6n - 5$
   d $10n - 1$
   e $3n + 8$
   f $\frac{1}{2}n + 1\frac{1}{2}$

CHAPTER 2 Number 1

Practice 2A Decimals

1. Without using a calculator, work out:
   a 12×10 b 68×100 c $83 \div 10$ d $62 \div 100$

2. Find the missing numbers:
   a $6 \div \boxed{} = 300$
   b $\boxed{} \div 100 = 0.06$
   c $600 \times \boxed{} = 60\,000$

3) Without using a calculator, work out:

a 1.3 × 10   b 0.06 × 10   c 0.7 × 100   d 6.4 × 1000
e 1.9 ÷ 10   f 2.9 ÷ 100   g 4.97 ÷ 100   h 5.81 ÷ 100

4) Find the missing numbers:

a 0.6 × 100 = ☐
b 6 ÷ ☐ = 0.6
c 0.6 ÷ 10 = ☐

5) Fill in the missing operation.

a 0.6 → ☐ → 0.06
b 0.65 → ☐ → 65
c 432 → ☐ → 4.32
d 0.265 → ☐ → 265

6) a Calculate the total cost of 100 plasters at £0.07 each.
   b Calculate the cost of this box of plugs.

PLUGS
1000 @ £0.18 each

c How much does one minidisk cost?

Mini disks £15

Practice 2B Ordering decimals

1) a Copy the table on page 16 of Pupil Book 2 (but not the numbers). Write the following numbers in the table, getting each digit in the appropriate column.

0.46, 0.09, 0.4, 0.38, 0.406, 0.309, 0.41

b Use the table from **a** to write the numbers in order from smallest to largest.

2) Write these sets of numbers in order from smallest to largest.

a 2.382, 1.893, 2.03, 2.4, 1.86
b 0.132, 0.031, 0.302, 0.123, 0.052

3) Put the correct sign > or < between these pairs of numbers.

a 3.62 … 3.26   b 0.07 … 0.073   c £0.09 … 10p

4 Put these amounts of money in order: £1.20   32p   £0.28   23p   £0.63

5 Put these lengths in order: 57 cm, 2.05 m, 0.06 m, 123 cm, 0.9 m

6 Write the weights of these cheeses in order. Hint: convert kilograms to grams.

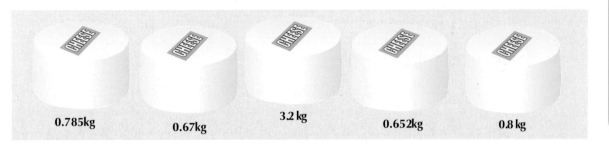

0.785kg   0.67kg   3.2 kg   0.652kg   0.8 kg

Practice 2C Directed numbers

1 Put the correct sign > or < between these pairs of numbers.

  a –6 … –4   **b** 5 … –4   **c** –2 … –8   **d** –5 … 3

2 Work out the answer to each of these.

  a –5 + 8   **b** 2 – 7   **c** –4 – 9   **d** 2 – 6 – 3
  e –5 + 6 – 4 **f** –3 + 8 – 2 **g** 5 – +7   **h** –7 + –3
  i 4 – –6   **j** – +7 – 4 + 6 **k** 2 – –5 – –8 **l** –4 + –6 – +3

3 Find the missing numbers to make these true.

  a –5 + ☐ = –3   **b** ☐ – 3 = –7

  c ☐ – –3 = 0   **d** –2 – ☐ = 6

4 In a magic square, the numbers in any row, column or diagonal add up to give the same answer. Could this be a magic square? Give a reason for your answer.

–3		–7
	4	
	–5	

Practice 2D Estimates

1 Explain why these calculations must be wrong.

  a 53 × 21 = 1111   **b** 58 × 34 = 2972   **c** 904 ÷ 14 = 36

2) Estimate answers to each of these problems.

  a 6832 − 496 b 28 × 123 c 521 ÷ 18

  d 770 × 770 e $\dfrac{58.9 + 36.4}{22.5}$

3) Which is the best estimate for 15.4 × 21.6?

  a 16 × 22 b 15 × 21 c 15 × 22 d 16 × 21

4) Estimate the number the arrow is pointing to.

5) **a** Football socks cost £3.71 a pair. Without calculating the answer, could Ian buy 5 pairs using a £20 note? Explain your answer clearly.
   b Shoelaces cost 68p a pair. The shopkeeper charges Ian £3.25 for 5 pairs. Without working out the correct answer, explain why this is incorrect.

Practice 2E Column method for addition and subtraction

1) By means of a drawing show how you would use a number line to work out the answers to these.

  a 4.7 + 3.8 b 9.3 − 3.16

2) Repeat the calculations in question 1 using column methods. Show all your working.

3) Use column methods to work out the following additions.

  a 19.8 + 23.1 b 28.4 + 6.92
  c 4.32 + 9.81 d 21.38 + 106.26 + 9.34

4) Use column methods to work out the following subtractions.

  a 17.6 − 9.4 b 28.6 − 12.93
  c 5.9 + 8.32 − 4.71 d 11.91 + 86.23 − 53.8

5) Work out the cost of a CD cleaning kit at £7.23, a CD carrying case at £5.69 and a sheet of CD labels at 84p.

6) A radio costs £26.48 and a CD player costs £52.13. How much dearer is the CD player?

Practice 2F Solving problems

1) 7 cans of beans weigh 1750 g. How much do 9 cans weigh?

2) An egg box containing 2 eggs weighs 140 g. The same egg box containing 3 eggs weighs 195 g. How much does an egg box containing 6 eggs weigh?

3. If 26 × 152 = 3952, write down, without calculating, the value of:

   a   2.6 × 152   b   2.6 × 15.2   c   260 × 1520

4. Find four consecutive even numbers that add up to 60.

5. To make a number chain, start with any number.
   When the number is even, divide it by 2 and subtract 1.
   When the number is odd, subtract 1 and double the answer.
   Investigate which numbers give the longest chains.

CHAPTER 3 Geometry and Measures 1

Practice

3A Length, perimeter and area

1. Measure the lengths of these pencils, giving your answers in centimetres.

   a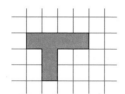

   b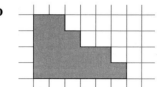

2. Find the perimeters of these shapes by using your ruler to measure the length of each side.

   a   b

   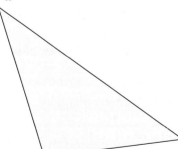   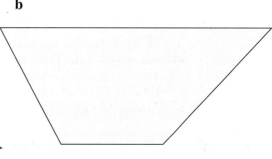

3. Copy these puzzle pieces onto 1 cm squared paper. Find the perimeter and area of each piece.

   a   b

4 Estimate the area of each of these shapes. Each square on the grid represents one square centimetre.

a 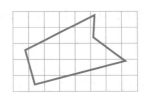 b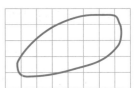

Practice 3B Perimeter and area of rectangles

1 i Find the perimeter of each rectangle.
  ii Find the area of each rectangle.

a
7 m, 16 m

b
22 mm, 9 mm

c
6 cm, 6 cm

d
3 mm, 28 mm

2 A swimming pool is 8 m wide and 30 m long.

   a Find the perimeter of the pool.
   b Find the area of the pool.
   c Emma wants to swim 1 kilometre.
      i How many widths does she need to swim?
      ii How many lengths does she need to swim?
   d The floor of the pool is covered with square tiles of side 50 cm. How many tiles cover the floor?

3 Use centimetre squared paper to draw three rectangles, each with area 24 cm².

4 i Find the perimeter of the following compound shapes.
  ii Find the area of the following compound shapes.

a 5 mm, 20 mm, 20 mm, 10 mm

b 12 cm, 9 cm, 5 cm, 3 cm

Practice 3C 3-D shapes

1 On squared paper, draw accurate nets for the following model storage boxes.

a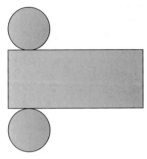

b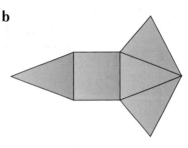

2 Below are the nets of some solids. Describe each solid.

a

b

3 Draw the following cuboid accurately on an isometric grid.

4 Use an isometric grid to draw solids made up of:

    a 4 cubes   **b** 10 cubes

Practice 3D Surface area of cubes and cuboids

1 Find the surface area for each of the following cubes.

a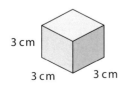

b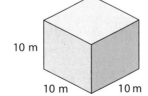

c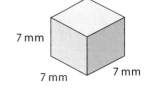

2 Find the surface area for each of the cubes with the following edge lengths.
   a 3 cm   **b** 6 cm   **c** 12 cm

3 Calculate the total surface area of the inside and outside of this jewel box, including the lid (ignore the thickness of wood).

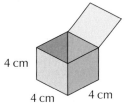

4 Five unit (centimetre) cubes are placed together to make these 3-D shapes.

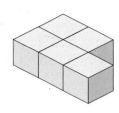

Find the surface area of each shape.

CHAPTER 4 Number 2

Practice

4A Fractions

1 Write down the fraction of each shape that is shaded.

a   b   c   d

2 Draw a fraction diagram to show each fraction.
   a $\frac{2}{3}$ **b** $\frac{1}{6}$ **c** $\frac{4}{5}$

3 Copy and complete the following equivalent fraction series.

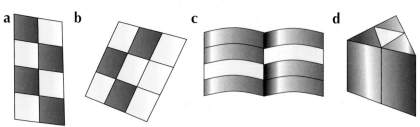

4 Find the missing number in each of these equivalent fractions.

a $\frac{3}{4} = \frac{\square}{16}$ b $\frac{4}{7} = \frac{20}{\square}$ c $\frac{7}{10} = \frac{\square}{70}$

5 Cancel these fractions to their simplest form.

a $\frac{5}{20}$ b $\frac{10}{12}$ c $\frac{21}{35}$ d $\frac{16}{20}$ e $\frac{12}{18}$ f $\frac{30}{25}$

6 The diagram shows the dial of a combination safe lock.

What fraction of a turn takes you from:
i A to C clockwise   ii A to C anticlockwise
iii H to E clockwise   iv J to D anticlockwise
v F to B anticlockwise   vi C to H clockwise?

7 a 1 day is 24 hours. What fraction of a day is 18 hours?
b 1 kilobyte is 1000 bytes. What fraction of a kilobyte is 450 bytes?

Practice

4B Fractions and decimals

1 Convert these top heavy (improper) fractions to mixed numbers.

a $\frac{5}{3}$ b $\frac{9}{2}$ c $\frac{10}{7}$ d $\frac{17}{4}$ e $\frac{26}{5}$ f $\frac{50}{9}$

2 Convert these mixed numbers to top heavy (improper) fractions.

a $1\frac{2}{3}$ b $3\frac{1}{8}$ c $2\frac{5}{6}$ d $5\frac{1}{4}$ e $9\frac{1}{2}$ f $6\frac{3}{10}$

3 Match the top-heavy fractions to mixed numbers:

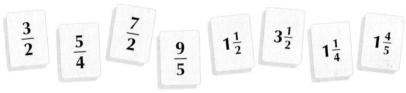

4 Convert the following decimals to fractions.

a 0.7   b 0.6   c 0.45   d 0.16   e 0.08   f 0.33

5 Convert the following fractions to decimals.

a $\frac{7}{10}$ b $\frac{19}{50}$ c $\frac{6}{25}$ d $\frac{13}{100}$ e $\frac{31}{25}$ f $\frac{9}{100}$

6 Put the correct sign < or > between these fractions.
Hint: Convert to fractions out of 100.

a $\frac{37}{100} \ldots \frac{17}{50}$ b $\frac{3}{10} \ldots \frac{11}{50}$ c $\frac{17}{25} \ldots \frac{7}{10}$

7 Put these fractions in order of size, smallest first. Use the hint in Question 6.

a $\frac{7}{20}, \frac{3}{10}, \frac{22}{50}$ b $\frac{18}{25}, \frac{3}{4}, \frac{71}{100}$ c $1\frac{7}{20}, 1\frac{1}{4}, 1\frac{13}{50}$

Practice 4C Adding and subtracting fractions

1 Work out each of the following:
   a $\frac{1}{7} + \frac{2}{7}$
   b $\frac{3}{11} + \frac{5}{11}$
   c $\frac{9}{11} - \frac{1}{11}$
   d $\frac{7}{9} - \frac{2}{9}$

2 Work out each of the following. Cancel down to lowest terms.
   a $\frac{3}{8} + \frac{3}{8}$
   b $\frac{7}{10} + \frac{1}{10}$
   c $\frac{8}{9} - \frac{2}{9}$
   d $\frac{11}{12} - \frac{5}{12}$

3 Add the following fractions.
   a $\frac{3}{8} + \frac{1}{4}$
   b $1\frac{1}{2} + \frac{5}{8}$
   c $2\frac{1}{4} + 1\frac{3}{8}$
   d $1\frac{1}{4} + \frac{7}{8} + 1\frac{1}{8}$

4 Add the following fractions.
   a $\frac{3}{7} + \frac{2}{7}$
   b $\frac{5}{8} + \frac{7}{8}$
   c $\frac{2}{11} + \frac{6}{11} + \frac{10}{11}$
   d $1\frac{3}{5} + \frac{4}{5}$

5 Subtract the following fractions.
   a $\frac{7}{8} - \frac{1}{4}$
   b $1\frac{1}{2} - \frac{5}{8}$
   c $2\frac{3}{8} - \frac{3}{4}$
   d $3\frac{5}{8} - 1\frac{3}{4}$

6 Subtract the following fractions.
   a $\frac{9}{10} - \frac{7}{10}$
   b $\frac{3}{8} - \frac{1}{8}$
   c $\frac{13}{16} - \frac{7}{16}$
   d $\frac{11}{8} - \frac{5}{8}$

7 Work along each chain of calculations. Write down your answer to each step.

   a $\frac{7}{10} \rightarrow +\frac{3}{10} = \boxed{} \rightarrow -\frac{4}{10} = \boxed{} \rightarrow +\frac{1}{10} = \boxed{} \rightarrow -\frac{6}{10} = \boxed{}$

   b $1\frac{3}{8} \rightarrow -\frac{5}{8} = \boxed{} \rightarrow +\frac{3}{4} = \boxed{} \rightarrow -\frac{3}{8} = \boxed{}$

   c $\frac{3}{8} \rightarrow +1\frac{3}{4} = \boxed{} \rightarrow +\frac{5}{8} = \boxed{} \rightarrow -1\frac{1}{2} = \boxed{} \rightarrow +2\frac{1}{8} = \boxed{} \rightarrow -\frac{7}{8} = \boxed{}$

Practice 4D Equivalences

1 Work out the following.
   a A quarter of thirty-two
   b A seventh of forty-nine
   c A tenth of two hundred and fifty

2 Work out the following.
   a $\frac{1}{4}$ of £60
   b $\frac{3}{5}$ of 30 kg
   c $\frac{4}{7}$ of 56 m
   d $\frac{8}{9}$ of 63p

3 Calculate the following.
   a 10% of £560
   b 80% of 160 m
   c 15% of 90 kg
   d 40% of 35 kg

4 Work out the equivalent percentages and fractions to the following decimals.
   a 0.7
   b 0.55
   c 0.96
   d 0.02

5 Work out the equivalent decimals and fractions to the following percentages.
  a 30%   b 16%   c 95%   d 5%

6 Work out the equivalent percentages and decimals to the following fractions.
  a $\frac{4}{10}$ b $\frac{17}{20}$ c $\frac{9}{25}$ d $\frac{37}{50}$ e $\frac{4}{5}$ f $\frac{5}{8}$

7 Find the pairs of equivalent numbers. Write your answers like this: **m = n**.

a 40%
b $\frac{3}{8}$
c 0.3
d 0.65
e 18%
f $\frac{2}{5}$
g 30%
h 0.375
i $\frac{13}{20}$
j $\frac{9}{50}$

Practice

4E Solving problems

1 A photocopier is set to enlarge images by 20%. What will be the new measurements when these objects are enlarged.
  a 20 cm
  b 150 mm
  c 8 cm

2 An improved assembly line reduces the time to assemble gamepads by 10%. What will the following assembly times reduce to?
  a 120 seconds   b 95 seconds   c 14 seconds

3 Which of these sales is the best deal?
  a SALE 35% off
  b SALE ⅖ off
  c SALE ¼ off

4 To calculate the tax paid on an annual salary, follow these steps:
  ▼ subtract £4500 from the salary (this amount is not taxed)
  ▼ calculate 10% of the next £1500 of the salary
  ▼ calculate 25% of the remaining salary.

Calculate the tax due for each of these salaries.
  a £6000   b £26 000   c £7000

5. Which of these is greater?

   a $\frac{3}{5}$ of 55 or $\frac{3}{4}$ of 48   b $\frac{5}{8}$ of 56 or $\frac{4}{7}$ of 49   c $\frac{5}{6}$ of 33 or $\frac{7}{10}$ of 35

6. Find one fraction that is between each pair of fractions.

   a $\frac{4}{25}, \frac{1}{5}$ b $\frac{11}{20}, \frac{6}{10}$ c $\frac{33}{50}, \frac{3}{4}$

7. A bag contains 240 jelly beans. $\frac{1}{6}$ are red, $\frac{3}{8}$ are yellow, $\frac{2}{5}$ are blue and the rest are green. How many of each colour are there? What fraction are green?

8. What number is halfway between the numbers shown on the following scales?

   a $\frac{1}{8}$ ———— $\frac{5}{8}$ b $\frac{5}{12}$ ———— $\frac{11}{12}$ c $1\frac{3}{8}$ ———— $2\frac{1}{8}$

CHAPTER 5 Statistics 1

Practice

5A Mode, median and range

1. Find the mode for each of the following sets of data.

   a win, draw, win, lose, lose, draw, win, draw, lose, lose, draw, win, draw
   b 5, 1, 7, 7, 3, 1, 5, 2, 6, 5, 3, 2, 4, 9

2. Find the range for each of the following sets of data.

   a 20, 70, 30, 25, 90, 35, 60, 60, 15
   b 3.1, 5.6, 2.9, 4.8, 3.6, 4.9, 6.3, 3.7

3. Find the median for each of the following sets of data.

   a 2, 7, 3, 12, 9, 15, 3
   b 31, 19, 17, 28, 40, 30, 42

4. Find the mode, median and range for each of the following sets of data.

   a 200 mm, 150 mm, 600 mm, 300 mm, 450 mm, 200 mm, 500 mm, 250 mm, 550 mm
   b £3.20 £2.90 £3.10 £3 £3.15 £2.90
   c 58 g, 48 g, 48 g, 52 g, 50 g, 51 g, 50 g, 56 g, 48 g, 58 g, 52 g, 56 g

5. 17 cars of the same year and model were tested for fuel efficiency. The table shows the miles per gallon (mpg) for the cars.

mpg	36	37	38	39	40
Number of cars (frequency)	3	4	7	2	1

a Calculate the mode.
b Calculate the median.
c Calculate the range.

Practice

5B The mean

1 Complete the following:

a mean of 2, 6, 7 = $\dfrac{2+6+7}{}$ = — = ☐

b mean of 5, 8, 11, 12 = $\dfrac{5+8+11+12}{}$ = — = ☐

c mean of 5, 5, 6, 9, 10 = $\dfrac{5+5+6+9+10}{}$ = — = ☐

d mean of 1, 3, 7, 17 = $\dfrac{1+3+7+17}{}$ = — = ☐

2 Complete the following:

a mean of 5, 9, 10 = ———— = — = ☐

b mean of 7, 11, 12 = ———— = — = ☐

c mean of 10, 14, 15, 17 = ———— = — = ☐

d mean of 18, 23, 27, 28 = ———— = — = ☐

3 Find the mean for each of the following sets of data.

a 3, 15, 7, 10, 23, 4, 9, 1
b 1.9, 4.2, 2.6, 3, 5.1, 1.8

4 Find the mean for each of the following sets of data, giving your answer to one decimal place.

a 3, 8, 4, 7, 9, 2, 4
b 53, 94, 21, 64, 100, 17, 35, 88, 42

5 The lap times, (in seconds) for seven players racing model cars are shown below.

20, 15, 16, 20, 15, 13, 20

a Find the mean lap time.
b Find the median lap time.
c Find the modal lap time.

Practice 5C Statistical diagrams

1. The dual bar chart shows car sales for two showrooms, Connors and VPW.

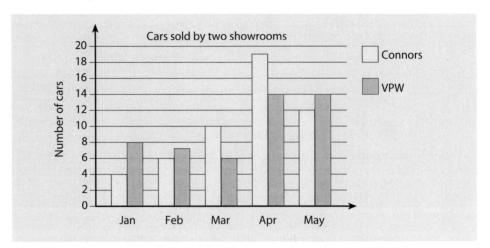

   a Which showroom sold the most cars?
   b During how many months did Connors sell more cars than VPW?
   c What were the most cars sold by each showroom in a month?
   d What is the range of cars sold for each showroom?

2. The bar chart below shows the number of batteries that a variety of torches need.

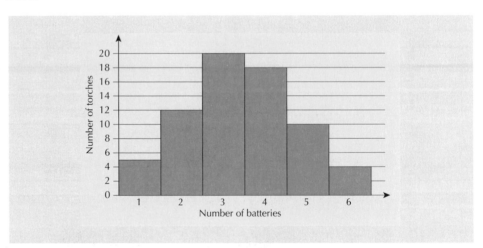

   a What is the mode?
   b What is the total number of torches sampled?
   c How many torches needed 4 batteries or less?

3. There are 240 trees in a small wood. The pie chart shows the proportion of each kind of tree.

   a Which kind of tree are there fewest of?
   b Estimate the fraction of pine trees.
   c Estimate the number of lime trees.

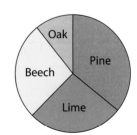

4 Jan checks the fuel in her motorbike tank at the end of each day. The line graph below shows the fuel levels for one week.

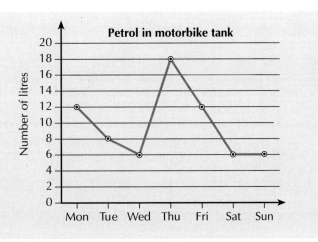

- **a** How much fuel was in the tank at the end of Friday?
- **b** When did the tank contain 6 litres of fuel?
- **c** On which day was the tank refilled?
- **d** What is the range of fuel in the tank for the week?
- **e** On which day did Jan not use her motorbike?

Practice 5D Probability

1 Choose one of the following words which best describes the probability for the events below.

impossible, very unlikely, unlikely, evens, likely, very likely, certain

- **a** A bonfire will cause smoke.
- **b** A telephone number ends with an odd digit.
- **c** A cup will break if it is dropped.
- **d** A 7 is thrown using an ordinary die.
- **e** A letter will get lost in the post.

2 Sweets contained in a bag of Lucky Numbers are shown below.

- **a** One sweet is picked from the bag, at random. What is the probability that the number is:
  i odd **ii** greater than 5 **iii** odd or even **iv** a multiple of 3?
- **b** Which number has a probability of $\frac{1}{5}$ of being picked?

3 Mark Choi has these chopsticks in his kitchen drawer.

Short	Long
6 red	2 red
2 green	8 green
4 yellow	2 yellow

He picks a chopstick at random. Find these probabilities.

a P(red)
b P(green or yellow)
c P(short)
d P(long and red)
e P(short and red **or** short and yellow)

4 Sema has made a Wheel of Fortune game. It costs 5p to play. The numbers on the wheel show the prizes, in pence.

If the wheel is spun once, what is the probability of winning the following prizes? Express your answers as decimals.

a 10p **b** nothing
c more than 5p **d** the biggest prize
e 7p

5 a Write down an event with a probability greater than 0 but less than $\frac{1}{2}$.
b Write down an event with a probability greater than $\frac{1}{2}$ but less than 1.

Practice

5E Experimental probability

1 These three coins were flipped together. Two of the coins are tails.

a Copy the tally chart below.

Result	Tally	Frequency
Two tails		
Other		

b Flip three coins 30 times. Record the results in the tally chart.

c Calculate the experimental probability of getting two tails.
d Is your answer greater or less than even chance?

2 a Trace this spinner and cut it out of thin card or paper. Push a drawing pin through the centre.
b Spin your spinner 50 times. Record your results in a frequency table.
c Calculate the experimental probability of the spinner landing on a letter.
d Is your answer greater or less than even chance?

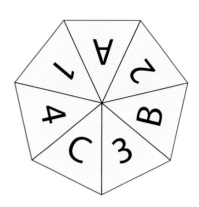

3 a Trace this spinner and cut it out of thin card or paper. Push a drawing pin through the centre.
b Spin your spinner 50 times and record your results in a copy of the frequency table below:

	Tally	Frequency
A		
B		
C		
1		
2		
3		
4		
Total:		

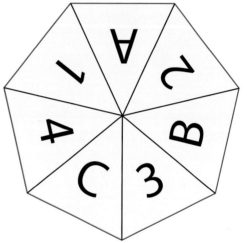

c Find your experimental probability of the spinner landing on a letter.
d Is your answer greater or less than an even chance?

CHAPTER 6 Algebra 2

Practice

6A Algebraic terms and expressions

1 Write terms, or expressions, to illustrate the following sentences.

a Subtract 2 from m.
b Divide x by 2.
c Multiply p by 7.
d Multiply t by 2, then subtract 1.
e Multiply c by c.
f Divide b by 10, then add 4.

2 Write down the values of the following terms for each different value of x.

    a $5x$ where   **i** $x = 3$ **ii** $x = 10$ **iii** $x = 100$
    b x^2 where   **i** $x = 4$ **ii** $x = 7$ **iii** $x = 9$
    c $\frac{x}{3}$ where   **i** $x = 6$ **ii** $x = 21$ **iii** $x = 30$
    d $0.5x$ where   **i** $x = 10$ **ii** $x = 8$ **iii** $x = 22$

3 Write down the values of these expressions for each different value of m.

    a $m - 3$ where   **i** $m = 4$ **ii** $m = 7$ **iii** $m = 15$
    b $m + 10$ where   **i** $m = 3$ **ii** $m = 10$ **iii** $m = 25$
    c $20 - m$ where   **i** $m = 2$ **ii** $m = 17$ **iii** $m = 20$
    d $3m + 2$ where   **i** $m = 2$ **ii** $m = 5$ **iii** $m = 9$
    e $100 - 5m$ where   **i** $m = 1$ **ii** $m = 7$ **iii** $m = 0$

4 Write down the values of these expressions for each different value of t.

    a $t^2 - 2$ where   **i** $t = 2$ **ii** $t = 7$ **iii** $t = 10$
    b $20 + t^2$ where   **i** $t = 5$ **ii** $t = 0$ **iii** $t = 9$

5 a Rearrange these cards to make five different expressions, for example, $3 + 2 - x$.

   b Calculate the value of each of your expressions when $x = 5$.

Practice
6B Rules of algebra

1 In each of the following clouds, only two expressions are equal to each other. Write down the equal pair.

        $4 + 7$ 7×4 $b - c$ $\frac{c}{b}$
  $7 \div 4$
           $4 \div 7$ $c + b$
    4×7 bc $c - b$
                            $b + c$ $\frac{b}{c}$

2 In each of the following lists, write down all the expressions that equal each other.

 a 2×5, $5 + 2$, $5 \div 2$, 5×2, $2 - 5$, $2 \div 5$, $2 + 5$, $5 - 2$

 b $\dfrac{t}{3}$, $t - 3$, $3t$, $t \div 3$, $3 + t$, $t \times 3$, $3 - t$

3 Write down two equivalent facts for each of the following facts.

 a $4 \times 6 = 24$ **b** $p - 5 = 7$
 c $10 + 3 = 13$ **d** $\dfrac{x}{2} = 3$

4 Show by substitution of suitable numbers that:

 a if $a - b = 3$, then $a = b + 3$ **b** if $pq = 20$, then $\dfrac{p}{20} = q$

5 Show that the following are true by using suitable substitutions.

 a $2b + c = c + 2b$ **b** $3ab = 3ba$

6 a^2 means $a \times a$ or simply aa.
 So $a^2 b$ means $a \times a \times b$ or simply aab.
 Write down three expressions equivalent to each of the following.

 a $a^2 b$ **b** $x^2 y^2$ **c** $10m^2$

Practice

6C Simplifying expressions

1 Simplify the following expressions.

 a $3p + 6p$ **b** $12q - 4q$ **c** $5t + 3t + 8t$
 d $4a + 3a - 7a$ **e** $9x - x$ **f** $20c - 6c - 6c$
 g $8m - m - 6m$ **h** $20k - 13k + 2k$ **i** $y + 17y - 9y$

2 Simplify the following expressions.

 a $2a + 3a + 5b + 6b$ **b** $5x + 3y + 3x + y$
 c $7d - 3d + 8e - 5e$ **d** $2 + 6m - 3m + 7$
 e $4p + 7q - 2p + 3q$ **f** $10t + 15 - 10 - 6t$
 g $5f + 7p - 2f + 3$ **h** $12k + 9m - 2m - 12k$
 i $9t + 9 - 3t - 7$

3 Expand the brackets.

 a $5(d + 3)$ **b** $2(5s + 3)$ **c** $10(a - 1)$
 d $4(6m - 3)$ **e** $20(a + b - c)$ **f** $4(3k - 2 + 4x)$
 g $6(10 - 2p - 3r)$

4 Expand and simplify the following expressions.

 a $2(x + 3) + 3(x + 5)$ **b** $4(m + 1) + 7(m + 2)$
 c $5(a + 2) + 2(a - 3)$ **d** $6(3d + 4) + 2(7d + 2)$
 e $2(3p + 8) + 3(2p - 1)$ **f** $8(3t + 4) + 7(2t - 3)$

Practice 6D Formulae

1 Write each of these rules as a formula.

    a The total distance, t, covered by running a number of laps, n, of a 400 m track.
    b The cost of one loaf of bread, b, if 5 loaves cost m.
    c The difference in height, d, between Sven, 180 cm tall, and Tammy, who is shorter. (Tammy is T cm tall.)
    d The total weight, w, of a box weighing 100 grams and an electric motor. (The motor weighs m grams.)

2 The formula shows the time taken (minutes) for a machinist to make a number of wooden chair legs.

$$t = 15 + 7n$$

where t = total time taken, n = number of chair legs. Find the time taken when:

    a $n = 3$ **b** $n = 50$ **c** $n = 17$

3 The number of workers it takes to complete a job is given by the formula below.

$$N = \frac{180}{t}$$

where N = number of workers, t = time spent by each person on the job. Find the number of people working on the job when:

    a $t = 60$ hours   **b** $t = 4$ hours   **c** $t = 15$ hours

4 The formula below gives the total cost of 15 plugs and sockets.

$$C = 15(P + S)$$

where C = total cost, P = price of a plug (in pence), S = cost of a socket (in pence).
Calculate the total cost when:

    a $P = 40p, S = 260p$ **b** $P = 25p, S = 100p$
    c $P = 32p, S = 172p$

5 The formula below gives the area of cardboard (cm²) needed to make a box of length 15 cm.

$$A = 15w - 4x$$

where A = area, w = width of box, x = side of square cut-out.
Calculate the area of cardboard needed when:

    a $w = 10$ cm, $x = 3$ cm   **b** $w = 20$ cm, $x = 5$ cm
    c $w = 17$ cm, $x = 4$ cm

Practice
6E Equations

1 Solve the following equations.

- **a** $5x = 20$
- **b** $10y = 10$
- **c** $8p = 32$
- **d** $9d = 72$
- **e** $2q = 300$
- **f** $7n = 154$
- **g** $3r = 81$
- **h** $4t = 240$

2 Solve the following equations.

- **a** $p + 5 = 12$
- **b** $m + 8 = 17$
- **c** $b + 30 = 100$
- **d** $h + 16 = 32$
- **e** $d - 3 = 8$
- **f** $p - 9 = 12$
- **g** $m - 23 = 17$
- **h** $s - 40 = 170$

3 Solve each of the following equations.

- **a** $\frac{x}{2} = 7$
- **b** $\frac{p}{2} = 9$
- **c** $\frac{y}{9} = 3$
- **d** $\frac{t}{20} = 9$

4 Solve each of the following equations.

- **a** $a + 6 = 14$
- **b** $d - 17 = 3$
- **c** $m + 8 = 3$
- **d** $s - 100 = 100$
- **e** $h - 10 = -3$
- **f** $q + 12 = -8$

5 Solve each of the following equations using number machines.

Example

$x + 3 = 51$

Working Write the equation using number machines:

$x \rightarrow \boxed{+3} \rightarrow 51$

$x \leftarrow \boxed{-3} \leftarrow 51$ (reverse the number machine)

$x = 51 - 3 = 48$ (work backwards to find the answer)

- **a** $y - 17 = 12$
- **b** $8x = 56$
- **c** $c + 23 = 18$

Chapter 7 Geometry and Measures 2

Practice 7A Lines and angles

1 Describe each of the following angles as acute, right-angle, obtuse or reflex. Estimate the size of each angle.

a   b   c

d   e   f

2 Write down four different properties of shape ABCDEF.

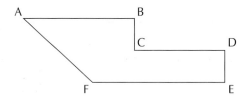

3 Write down the geometric properties of this kite.

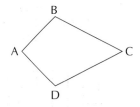

4 For each question, mark any equal sides, parallel sides, right-angles, etc.

   a Sketch a quadrilateral with exactly two parallel lines and two pairs of perpendicular sides.
   b Sketch a pentagon (five-sided shape) with exactly three equal sides and one right angle.
   c Sketch a hexagon (six-sided shape) with exactly two parallel lines, three obtuse angles and two acute angles.

Practice 7B Calculating angles

Calculate the sizes of the missing angles.

1
a
b
c
d

2
a
b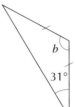
c
d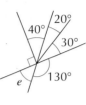

3
a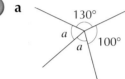
b

Practice 7C Coordinates

1 Write down the coordinates of the points A, B, C, D and E.

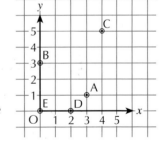

2
a Make a copy of the grid in question 1 and plot the points A(1, 1), B(3, 1) and C(5, 4).
b The three points are the vertices of a parallelogram. Plot the point D to complete the parallelogram.

3 Write down the coordinates of the points A, B, C, D, E, F and G.

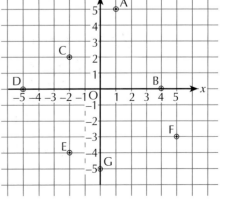

4
a Make a copy of the grid in question 3 and plot the points A(−5, −3), B(−5, 4) and C(2, 4).
b Join the points to make a triangle.
c Plot point D so that ABCD is a square. Draw the other two sides of the square.
d Each diagonal of the square crosses the axes at two points. Write down their coordinates.

CHAPTER 8 Statistics 2

Practice 8A Using a tally chart

Local residents were asked how they would like to see some waste ground developed. The results are shown in the tally chart below.

Development	Tally	Frequency
Swimming pool	|||| /	
Playground	|||| |||| //	
Supermarket	|||| ////	
Health centre	||||	
Cafe	///	

1. Draw a bar chart to illustrate the data.

2. Write possible reasons why the residents voted for each kind of development.

Practice 8B Using the correct data

You are going to investigate the number of words per line in a paperback novel. Follow these steps:

1. Copy this tally chart. You will need 10 to 15 rows.

Number of words	Tally	Frequency
1		
2		
3		
4		

2. Find a paperback novel. Choose a page at random. Count the number of words on each line. Do this for 50 lines. Record the data in your tally chart.

3. Decide how to count hyphenated words, numbers etc.

4 Draw a bar chart to illustrate your data.

5 If you have time, repeat the activity using a different paperback. Compare your results.

Practice

8C Grouped frequency

1 Some pupils were asked how many music CDs they owned. The results are shown below.

14	1	4	7	0	11	7	2	17	9
20	6	10	0	5	3	21	0	15	6
11	0	13	3	8	2	18	9	27	1
12	8	1	18						

a Copy and complete this grouped frequency table.

Number of CDs	Tally	Frequency
0–4		
5–9		

b Draw a bar chart to illustrate the data.

2 The average ages of families at a small holiday resort are shown below.

31	17	31	21	32	29	27	39	25	25
40	27	36	28	46	19	38	32	23	28
35	19	41	30	24	34	55	26	20	36
43	51								

a Which is the most sensible table for the data? Explain your answer.

Average age	Tally	Frequency
10–14		
15–20		

Average age	Tally	Frequency
0–19		
20–39		

Average age	Tally	Frequency
10–19		
20–29		

b Copy and complete the table you chose in part **a**.
c Draw a bar chart to illustrate the data.

8D Data collection

Kickoff Ltd closes down during the summer holiday. Office and factory staff were asked these questions about their preferred holiday times.

'How many days of your annual leave would you like to use for your summer holiday?'
'In which month would you like your summer holiday?'
'On which day of the week should the holiday begin?'

Their answers are shown in the table below.

Employee	Male/Female	Days of annual leave	Month	Weekday
Office	M	10	June	Monday
Office	F	5	August	Monday
Factory	M	5	August	Wednesday
Office	M	10	June	Monday
Factory	F	8	July	Wednesday
Factory	M	12	August	Thursday
Factory	M	10	July	Monday
Office	F	10	August	Monday
Factory	F	5	July	Wednesday
Factory	F	15	July	Monday
Factory	M	10	June	Monday
Factory	F	5	August	Monday
Factory	M	10	August	Monday
Office	F	12	July	Thursday
Factory	F	10	August	Monday
Factory	M	5	August	Monday
Factory	F	5	July	Friday
Office	F	10	June	Monday
Office	M	5	August	Wednesday
Factory	F	10	July	Monday

1 a Make a frequency table for the number of days chosen by office staff.
   b Make a frequency table for the number of days chosen by factory staff.
   c Write a sentence comparing office and factory staff.

2 a Make a frequency table for the holiday months chosen by male staff.
   b Make a frequency table for the holiday months chosen by female staff.
   c Write a sentence comparing male and female staff.

3 a Make a frequency table for the start day chosen by office staff.
   b Make a frequency table for the start day chosen by factory staff.
   c Write a sentence comparing office and factory staff.

 # Number and Measures 3

Practice

9A Rounding

1) Round off these numbers to:
  i the nearest 10   **ii** the nearest 100   **iii** the nearest 1000

  a 657   **b** 2555   **c** 3945   **d** 409   **e** 17 059   **f** 9895

2) **i** What are these Test Your Strength scores, to the nearest 100?
  ii Estimate the scores to the nearest 10.

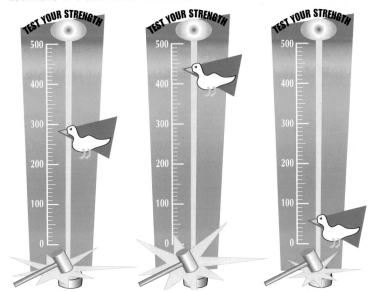

3) **i** Read these weighing scales, to the nearest gram.
  ii Read the scales to the nearest 10 g.
  iii Read the scales, correct to one decimal place

grams

grams

4) Round off these numbers to:
  i the nearest whole number   **ii** one decimal place

  a 7.32   **b** 8.75   **c** 3.04   **d** 19.58   **e** 0.749   **f** 9.955

5) Round these bank balances using a sensible degree of accuracy. Then place them in order of size, starting with the smallest.

Andrew	£4960.78	Jamil	£680	Geoff	£4799.50
Jean	£12 600	Suzy	£6709.67	Laurence	£94.45
Harry	£6921.44	Sema	£4934.24	Darren	£5094.68

Practice 9B The four operations

Show your working for each question.

1. Calculate the total cost of 35 burger meals at £3.98 each.

2. Electric light bulbs can be packed into boxes of 16 or 24. How can 232 bulbs be packed into full boxes only? Find two possible ways.

3. Which is better value?

4. **a** Find two consecutive even numbers whose product is 288.

   b Find three consecutive odd numbers whose sum is 111.

5. A postmaster has a full sheet of 80 first-class stamps costing 27p each. He also has a full sheet of 90 second-class stamps costing 19p each.

   a What is the total value of the two sheets of stamps?
   b He sells 31 first-class stamps. What is the value of the rest of that sheet?
   c How many second-class stamps can be bought for £5?
   d How many first-class stamps can be bought for £10?

6. A full bottle contains between 80 ml and 90 ml of medicine. If it were used for 5 ml doses, there would be 3 ml left in the bottle. If it were used for 3 ml doses, there would be 1 ml left in the bottle. How much medicine is in a full bottle?

Practice 9C BODMAS

1. Circle the operation that you do first in these calculations, then work it out.

   a $9 - 6 \div 3$ **b** $3 \times (20 - 15)$ **c** $50 \div 10 - 5$
   d $(3 + 13) \div 4$

2. Work out the following, showing each step of the calculation.

   a $20 + 20 \div 4$ **b** $4 \times 8 - 3 \times 7$ **c** $(8 - 2) \times 4$
   d $6 + 4^2$

3 Put brackets into the following to make the calculation true.

a $20 - 10 - 3 = 13$
b $5 + 2 \times 2 = 14$
c $9 - 7 - 5 - 3 = 4$
d $5 - 2^2 = 9$
e $12 - 3 \times 10 - 6 = 0$
f $10 + 10 \div 10 + 10 = 1$
g $1 + 3^2 + 5^2 = 41$
h $40 - 20 \div 4 + 1 = 36$

4 Work out the value of the following.

a $5^2 - 5$
b $(5 + 5) \div 5 + 5$
c $5 - 5 \times (5 - 5)$
d $5 \times 5 - (5 \div 5) \times (5 + 5)$

5 Write down each of the following using a single calculation. Use numbers and symbols $+, -, \times, \div, (\ )$ only. Then calculate the answer.

a Subtract the product of 4 and 2 from 12.
b Multiply the sum of 9 and 8 by 3.
c Add the square of 3 to 5.
d Subtract the square of 3 from the square of 4. Then subtract the answer from 20.

Practice
9D Long multiplication and long division

1 Work out the following long multiplication problems. Use any method you are happy with.

a 13×32
b 54×27
c 19×275
d 148×38

2 Work out the following long division problems. Use any method you are happy with. Some of the problems will have a remainder.

a $420 \div 12$
b $600 \div 22$
c $738 \div 38$
d $884 \div 26$

Decide whether these problems are long multiplication or long division. Then do the appropriate calculation, showing your method clearly.

3 A sports stadium has 23 rows of seats. Each row has 84 seats. How many people can be seated in the stadium?

4 A bag contains 448 g of flour. The flour is used to make cakes. Each cake contains 14 g of flour.

a How many cakes were made?
b How much flour is needed to make 234 cakes?

5 Floor tiles measure 35 cm by 26 cm. They cover a floor measuring 945 cm by 650 cm.

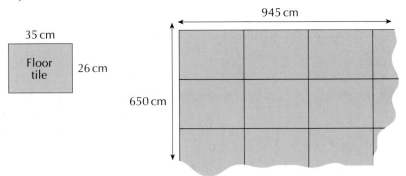

a How many tiles are next to the long edge of the floor?
b How many tiles are next to the short edge of the floor?
c How many tiles were used to cover the floor?

9E Efficient calculations

1 Without using a calculator work out the value of the following.

a $\dfrac{20 - 6}{2 + 5}$ b $\dfrac{7 + 9}{0.9 - 0.4}$

2 Use a calculator to do the calculations in question 1.

Did you get the same answer as before?

3 For each part of question 1, write down the sequence of keys that you pressed.

4 Work out the value of the following. Round off your answers to 1 decimal place, if necessary.

a $\dfrac{689 + 655}{100 - 58}$ b $\dfrac{420 - 78}{54 \div 3}$ c $\dfrac{36 \times 84}{29 + 17}$ d $\dfrac{296 + 112}{183 - 159}$

5 a Estimate the answer to $\dfrac{443 + 178}{53 - 27}$.

b Now use a calculator to work out the answer to 1 decimal place. Is your answer about the same?

6 Calculate the following.

a $\sqrt{344\,569}$ b 5.4^2
c $\sqrt{(50.6 + 39.65)}$ d $(12.3 - 2.6)^2$

7 a Calculate $\dfrac{387 + 231}{306 \div 17}$.

b Write your answer to part **a** as a mixed number.

Practice 9F Calculating with measurements

1 Convert the following measurements.

- a 160 mm to cm
- b 0.53 m to cm
- c 0.035 km to m
- d 9.7 m to cm
- e 5200 m to km
- f 37 m to km
- g 4000 cm to km
- h 23 cm to mm
- i 8.25 m to mm
- j 9 km to cm
- k 3400 g to kg
- l 9.32 kg to g
- m 265 g to kg
- n 0.01 kg to g
- o 9 g to kg
- p 3200 ml to l
- q 5.729 l to ml
- r 115 minutes to hours/minutes
- s 924 minutes to hours/minutes

2 Add together the following measurements and give the answer in an appropriate unit.

- a 0.063 kg, 580 g, 0.4 kg
- b 450 ml, 0.63 l, 9 cl
- c 640 mm, 94 cm, 0.003 km

FM 3 Fill in the missing unit.

- a A car weighs 1262 …
- b A £1 coin is about 3 … thick.
- c A glass holds about 30 … of milk.
- d Scott ran 200 … in a time of 32 …

CHAPTER 10 Algebra 3

Practice 10A Square numbers and square roots

1 Write down the value of the following. (Do not use a calculator.)

- a $\sqrt{64}$
- b $\sqrt{1}$
- c $\sqrt{121}$
- d $\sqrt{81}$
- e $\sqrt{0}$

2 With the aid of a calculator, write down the value of the following.

- a $\sqrt{196}$
- b $\sqrt{784}$
- c $\sqrt{8281}$
- d $\sqrt{344\,569}$
- e $\sqrt{3\,364\,000\,000}$

3 Make an estimate of the following square roots, then use a calculator to see how many you got right.

- a $\sqrt{169}$
- b $\sqrt{484}$
- c $\sqrt{900}$
- d $\sqrt{3600}$
- e $\sqrt{2209}$

4 Sometimes, the difference between two square numbers is another square number. For example,
$10^2 - 8^2 = 100 - 64 = 36$ and $6^2 = 36$, so $10^2 - 8^2 = 6^2$.

Use the numbers in the cloud to find more of these. Write each answer like this: $10^2 - 8^2 = 6^2$.

```
        4
     3    7
  10   15
5    8  24 12  9
          13  25
   20  16   26
      6
```

5 a Copy and continue the pattern to make eight rows. Work out all of the answers.

$1 \quad\quad\quad =$
$1 + 3 \quad\quad =$
$1 + 3 + 5 \quad =$
$1 + 3 + 5 + \ldots \quad =$

b What can you say about the answers?
c Can you find a rule that gives the answer? Check that your rule works.

Practice
10B Triangle numbers

1 Look at the numbers in the box. Write down the numbers from the box that are:

a square numbers **b** triangle numbers
c odd numbers **d** multiples of 3
e factors of 48 **f** prime numbers

1, 2, 3, 4, 5, 6, 13, 16, 18, 21, 22, 36, 40, 41, 45

2 Each of the following numbers is the sum of two triangle numbers. Write them down, for example, $16 = 1 + 15$.

a 4 **b** 18 **c** 38 **d** 42 **e** 70

3 Every odd number can be written as the difference between two triangle numbers, for example, $9 = 15 - 6$.

Express the following odd numbers as the difference between two triangle numbers.

a 5 **b** 11 **c** 21 **d** 29

4 Write down the first 20 triangle numbers. Cross out the numbers in these positions: 2nd and 3rd, 5th and 6th, 8th and 9th and so on. Calculate the differences for the remaining sequence. Describe the differences.

Practice

10C From mappings to graphs

1 For each of the following:
   i   complete the input/output diagram.
   ii complete the coordinates alongside.
   iii plot the coordinates and draw the graph.

a × 4   **Coordinates**

   0 → 0   (0, 0)
   1 → 4   (1, 4)
   2 →   (2,)
   3 →   (3,)
   4 →   (4,)
   5 →   (5,)

b + 4   **Coordinates**

   0 → 4   (0, 4)
   1 → 5   (1, 5)
   2 →   (2,)
   3 →   (3,)
   4 →   (4,)
   5 →   (5,)

c − 1   **Coordinates**

   1 → 0   (1, 0)
   2 →   (2,)
   3 →   (3,)
   4 →   (4,)
   5 →   (5,)

Practice 10D Naming graphs

1 Write down the name of the straight line that passes through each of these pairs of points.

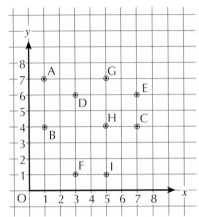

- **a** A and G
- **b** E and C
- **c** B and H
- **d** D and F
- **e** C and E
- **f** F and I

2 Draw the following graphs on the same grid. Label each line.

Use these axes:
x-axis from 0 to 8
y-axis from 0 to 8

- **a** $y = 2$
- **b** $x = 2$
- **c** $y = 8$
- **d** $x = 6$
- **e** $y = 5$
- **f** $x = 7$

3 Write down the letters which lie on the following lines.

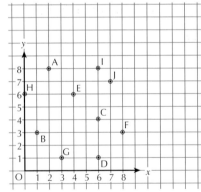

- **a** $x = 4$
- **b** $y = 3$
- **c** $x = 2$
- **d** $y = 6$
- **e** $x = 8$
- **f** $y = 1$

Practice 10E Naming sloping lines

1 Copy and complete the following input/output diagram for each of the stated functions.

→ function →	Coordinates
2 →	(2,)
3 →	(3,)
4 →	(4,)
5 →	(5,)
6 →	(6,)

a → +1 →   b → −1 →   c → +2 →   d → −2 →

2 a Plot each set of coordinates from Question 1 on the same axes.

Use these axes:
x-axis from 0 to 7
y-axis from 0 to 7

b Match each line to one of the following equations.
$y = x - 1$ $y = x + 2$ $y = x - 2$ $y = x + 1$

3 Answer these questions without drawing graphs.
a Is the point (3, 6) on the graph of $y = 2x$?
b Is the point (20, 16) on the graph of $y = x - 6$?
c Is the point (5, 17) on the graph of $y = 3x + 2$?
d Is the point (9, 76) on the graph of $y = 10x - 14$?

4 Which of the following lines does the point (5, 9) lie on?
$y = 4x$ $y = x + 3$ $y = 2x - 1$ $y = 5$

Chapter 11
Geometry and Measures 3

Practice 11A Measuring and drawing angles

1. Measure the size of each of the following angles, giving your answers to the nearest degree.

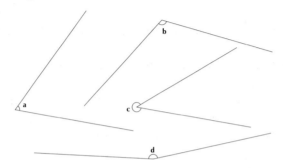

2. Draw and label the following angles.

   a 70°   b 42°   c 134°   d 200°   e 343°

3. a Measure all the angles in the quadrilateral ABCD.
   b Add the angles together.
   c Comment on your answer.

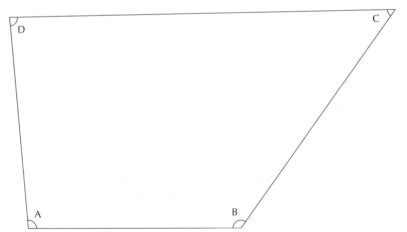

Practice 11B Constructions

1 Construct the following triangles. Remember to label the vertices and angles.

a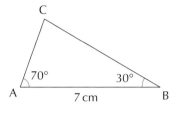
b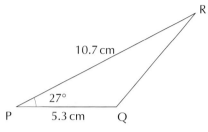

  c On your drawing, measure sides BC and AC, to the nearest millimetre.
  d On your drawing, measure ∠Q and ∠R, to the nearest degree.

2 a Construct the triangle ABC with AB = 6.4 cm, BC = 11.3 cm, ∠B = 16°.
  b Measure side AC, to the nearest millimetre.
  c Measure angles ∠A and ∠C, to the nearest degree.

3 a Construct the quadrilateral PQRS.
  b On your drawing, measure ∠R and ∠S, to the nearest degree.
  c On your drawing, measure RS, to the nearest millimetre.

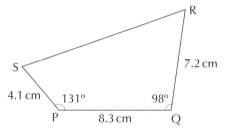

Practice 11C Solving geometrical problems

1 Draw the different types of quadrilateral, for example, square, rectangle, etc. Draw one diagonal for each quadrilateral. Describe the triangles you have made.

2 Explain the difference between a parallelogram and a trapezium.

3 How many different kinds of quadrilateral can be constructed on this pin-board?

Use square dotted paper to record your quadrilaterals. Beneath each diagram, write the name of the quadrilateral.

4 A square is a special kind of rectangle. Fill in the blanks to make different statements.

  a A parallelogram is a special kind of …
  b A … is a special kind of parallelogram.
  c A square is a special kind of …
  d A … is a special kind of kite.

CHAPTER 12 Number 4

Practice

12A Percentages

1. Write each percentage as a combination of simple percentages.
   a 21%   b 32%   c 76%   d 95%

2. Without using a calculator, work out each of these.
   a 20% of 70   b 21% of 40   c 32% of 24   d 76% of 400

3. Work out each of these using a calculator.
   a 95% of 220 kg   b 31% of 6 m   c 53% of 72 years
   d 11% of 17 litres   e 1% of 723 pages   f 99% of 634 houses
   g 6% of 4200 seeds   h 73% of £2427

4. Which is bigger?
   a 19% of £73 **or** 32% of £56   b 9% of 523 g **or** 99% of 46 g

5. Write down or work out the equivalent percentage and decimal to each of these fractions.
   a $\frac{3}{4}$ b $\frac{4}{5}$ c $\frac{1}{20}$
   d $\frac{17}{20}$ e $\frac{1}{25}$ f $\frac{9}{25}$
   g $\frac{1}{8}$ h $\frac{7}{8}$

6. Write down or work out the equivalent percentage and fraction to each of these decimals.
   a 0.4   b 0.9   c 0.27   d 0.65   e 0.62

7. Write down or work out the equivalent fraction and decimal to each of these percentages.
   a 35%   b 60%   c 28%   d 17.5%   e 43%

8. Arrange these numbers in increasing order of size.
   Hint: Convert the numbers to percentages.
   a 0.65, $\frac{16}{25}$, 63%   b 0.09, $\frac{1}{8}$, 11%

Practice 12B Ratio and proportion

1 5 reels of fishing line have a total length of 72 m. What is the total length of:
 a 1 reel? b 3 reels? c 11 reels?

2 The table shows the seeds contained in four packets of Summer Border.

Flower	Number of seeds
Cloth of Gold	20
Delphinium	12
Penstemon	8
Mrs Perry	6

Calculate the number of each seed contained in:

 a 8 packets   b 6 packets   c 10 packets

3 For each of these shapes, work out the following.
 i What proportion is shaded?
 ii What is the ratio of the shaded part to the unshaded part?

 a 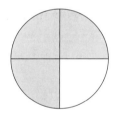   b 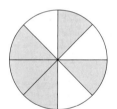   c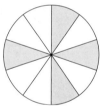

4 The contents of two boxes of muesli are shown below.

a What is the total weight of each box?
b Copy and complete the table below, which shows the proportion of each ingredient.

	Oats	Wheat	Nuts	Fruit
Luxury Muesli	25%			
Natural Breakfast				

5 In an orchestra, the ratio of cellists to violinists is 3 : 5. There are 20 violinists in the orchestra. How many cellists are there?

Practice

12C Calculating ratios and proportions

1 Reduce each of the following ratios to its simplest form.
a 9 : 6   b 8 : 20   c 24 : 30   d 200 : 900
e 75 : 45   f 21 : 56

2 Write down the ratio of coloured : white squares for these grids.

a 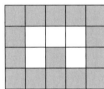   b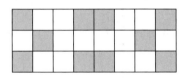

3 Four football teams scored 250 goals altogether. Beetwood Rangers scored 50 goals, Trenton FC 35, Middleton Wanderers 90 and Pinkhams scored the rest.

a Write down the percentage of the total goals that each team scored.
b Write down each of the following goal ratios in its simplest form.
 i Beetwood Rangers : Trenton FC
 ii Middleton Wanderers : Pinkhams
 iii Pinkhams : Beetwood Rangers

4 The atmosphere of Uranus is made up of Hydrogen and Other Gases in the ratio 6 : 1. A space probe filled two flasks with samples of the atmosphere.

a One flask contained 18 litres of Hydrogen. How much Other Gases did it contain?
b One flask contained 0.8 litres of Other Gases. How much Hydrogen did it contain?

Practice 12D Solving problems

1 Divide £180 in each of these ratios.

    a 4 : 5   **b** 11 : 4   **c** 5 : 7   **d** 22 : 23

2 There are 96 houses on an estate. The ratio of detached to semi-detached houses is 11 : 5. How many of each kind of house are there?

3 Andrea divides up her free time in the ratio:

    sewing : reading : photography = 7 : 6 : 12

She has 300 minutes of free time. How much time does she spend on each activity?

4 Dominic has 30 coloured sweets left in a packet. These are in the ratio:

    red : green = 2 : 3

    a How many of each colour are there?
    b If he eats 2 red sweets and 3 green sweets, what is the new ratio red : green? Comment on your answer.

5 Share 400 kg in each of these ratios.

    a 1 : 2 : 5   **b** 6 : 11 : 3

CHAPTER 13 Algebra 4

Practice 13A Solving 'brick wall' problems

Find the unknown number x in each of these 'brick wall' problems.

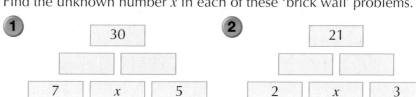

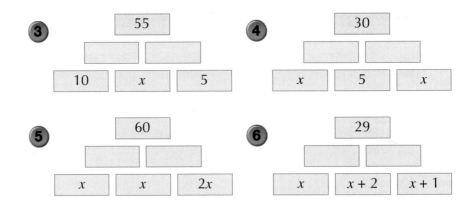

Practice
13B Solving square-and-circle problems

Find the solution set to each of the following square-and-circle puzzles. All solutions must use positive numbers.

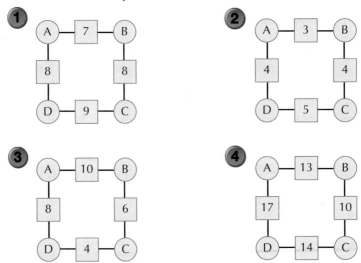

Practice
13C Triangle-and-circle problems

1. Use algebra to solve the triangle-and-circle problems. All solutions use positive integers.

a
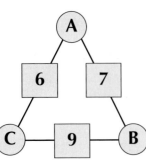

b

c
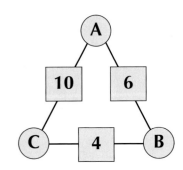

2 Use algebra to solve the triangle-and-circle problems. The solutions involve positive and negative integers.

a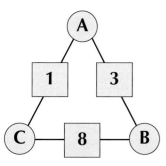

b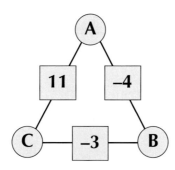

c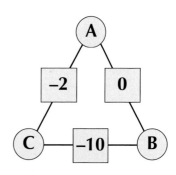

CHAPTER 14 Geometry and Measures 4

Practice 14A Line symmetry

1 Copy each of these shapes and draw its lines of symmetry. Write below each shape the number of lines of symmetry it has.

a B

b W

c

d

e

2 Write down the number of lines of symmetry for each of the following shapes.

a

b

c

d

3 a Copy each shape onto squared paper. Draw an extra line to make the shape have one line of symmetry.

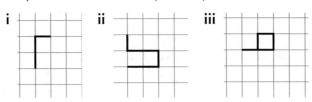

b Copy each shape onto squared paper. Draw extra lines to make the shape have two lines of symmetry.

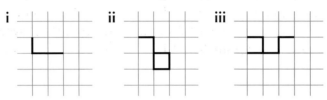

Practice
14B Rotational symmetry

1 Copy each of these shapes. Write the order of rotational symmetry beneath the shape. Mark the centre of rotation.

a   b   c   d

2 Write down the order of rotational symmetry for each of the following shapes.

a   b   c   d

3 a Copy each shape onto squared paper. Shade an extra square to give the shape rotational symmetry, order 2. Mark the centre of rotation.

i 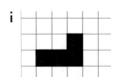 ii 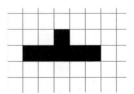 iii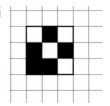

b Copy each shape onto squared paper. Shade extra squares to give the shape rotational symmetry, order 4. Mark the centre of rotation.

i 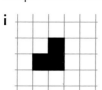 **ii** 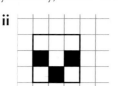 **iii**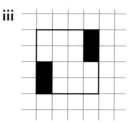

4 Draw new shapes on squared paper with the following properties.

    a Rotational symmetry, order 4.
    b Rotational symmetry, order 2, with no line symmetry.
    c Rotational symmetry, order 2, with two lines of symmetry.
    d Rotational symmetry, order 4, with four lines of symmetry.

Practice

14C Reflections

1 Copy each of these diagrams onto squared paper and draw its reflection in the given mirror line.

a 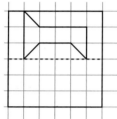   **b**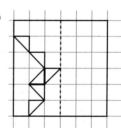

c 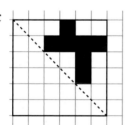   **d**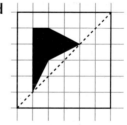

2 Look at the points shown on the grid below.

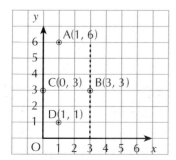

a Copy the grid onto squared paper and plot the points A, B, C and D. Include the coordinates on your diagram. Draw the mirror line.
b Reflect the points in the mirror line and label them A', B', C' and D'.
c Write down the coordinates of the image points.
d The point E' (4, 5) is the reflection of point E. Mark both points E and E' on your diagram, including the coordinates.

3 The diagrams show the reflections of numbers (images). Copy each diagram. Draw the numbers (objects).

a ƐƧϱ

b ⊥8ϱ

c ᄅ⌒⊥

Practice 14D Rotations

1 Copy each of the shapes below onto a square grid. Draw the image after each one has been rotated about the point marked X through the angle indicated. Use tracing paper to help.

a 90° clockwise

b 180°

c 90° anti-clockwise

d 90° anti-clockwise

2 **a** Rotate the triangle ABC through 90° anti-clockwise about the point (4, 4) to give the image A'B'C'.
b Write down the coordinates of A', B' and C'.
c Which coordinate point remains fixed throughout the rotation?
d Fully describe the rotation that will map the triangle A'B'C' onto the triangle ABC.

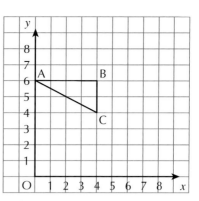

3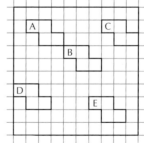

90° clockwise   180°   90° anti-clockwise

a Copy each number.
b Rotate the number through the given angle. Use the dot as the centre of rotation.

Practice

14E Translations

1 Describe each of the following translations.

  a A to C
  b A to D
  c C to B
  d D to E
  e B to A
  f B to D
  g D to C

2 Copy the grid and kite ABCD onto squared paper. Label it M.

  a Write down the coordinates of the vertices of kite M.
  b Translate kite M 2 units left, 6 units down. Label the new kite P.
  c Write down the coordinates of the vertices of kite P.
  d Translate kite P 5 units left, 6 units up. Label the new kite Q.
  e Write down the coordinates of the vertices of kite Q.
  f Translate kite Q 1 unit right, 3 units down. Label the new kite R.
  g Write down the coordinates of the vertices of kite R.
  h Describe the translation that maps kite R onto kite S.

CHAPTER 15 Statistics 3

Practice 15A Pie charts

1. This pie chart shows the percentage of people who voted for their favourite computer game.

   What percentage of the people voted for:
   a   Gears of War   b   Pro Soccer
   c   Sims   d   Sonic?

2. The pie chart on the right shows the percentage contribution to air pollution for 2000.

   a   What percentage of the world's air pollution in 2000 was from:
   i   Europe   ii   Asia
   iii Africa   iv   America?

   The pie chart on the right shows the percentage contribution to air pollution for 2006:

   b   What percentage of the world's air pollution in 2006 was from:
   i   Europe   ii   Asia   iii   Africa
   iv   America?
   c   Which areas of the world decreased their proportion of world air pollution?
   d   The proportion of air pollution stayed the same for one area of the world. Which area?

Practice 15B Comparing data

1. The data below shows pizza delivery times, in minutes, for two rival companies.

   Pizza-to-Go   16, 22, 18, 19, 22, 17, 17, 19, 19, 21, 18, 20
   Whizza Pizza 14, 20, 13, 16, 15, 27, 19, 22, 12, 15, 17, 14

   a   Calculate the mean delivery time for Pizza-to-Go.
   b   Calculate the mean delivery time for Whizza Pizza.
   c   Calculate the range for Pizza-to-Go.
   d   Calculate the range for Whizza Pizza.
   e   Which company would you use?

2 Class 7A has to decide which person will represent them in the annual School Brain competition. The results of the top two children in a range of tests are shown below.

James	7, 9, 6, 7, 5, 1, 8, 10
Parminder	5, 7, 8, 6, 5, 8, 6, 9

  a Calculate the mean score for James.
  b Calculate the mean score for Parminder.
  c The person with the highest mean was chosen to represent the class. Explain why this might not be the best person.

3 Two brands of carrot seeds were compared. The carrot harvest, in kilograms, from 10 packets of each brand are shown below.

Studley Seeds	12, 7, 11, 5, 13, 8, 7, 7, 11, 10
Super Seeds	9, 11, 8, 8, 9, 9, 11, 12, 8, 9

Which brand of seed do you think is best?

Practice

15C Statistical surveys

1 a Find a sample of 30 people or more and show them the number 4719352 for ten seconds, then ask them to remember it.

After ten seconds, take the number away, wait ten seconds, then ask them what the number was. They must remember it exactly to count as 'remembered'. Use the data to complete the following tally chart.

Remembered?	Tally	Frequency
Yes		
No		

  b Would you say that most people can remember a 7-digit number for a short time?

2 Find a typical weekend's football results (either English, Scottish or Australian).

  a Use the data to complete the following tally chart.

Number of goals scored by a team	Tally	Frequency
0		
1		
2		
3		
4		
5		
More than 5		

b Illustrate your results with a bar chart.
c From your data, what would be the typical number of goals scored by a football team?

Practice 15D Probabilities from two way tables

1 Lauren did a survey about her class walking to school or not:

	Number of boys	Number of girls
Walk to school	10	9
Do not walk to school	3	7

   a How many pupils are in Lauren's class?
   b How many boys in Lauren's class walk to school?
   c How many girls are there in Lauren's class?
   d What is the probability of selecting a pupil at random that:
      i is a boy who walks to school?
      ii is a girl?

2 Anna did a survey of how many of her class have Sky TV:

	Number of boys	Number of girls
Have Sky TV	11	4
Do not have Sky TV	5	8

   a How many pupils are in Anna's class?
   b How many girls in Anna's class have Sky TV?
   c How many boys are there in Anna's class?
   d What is the probability of selecting a pupil at random that
      i is a girl who has Sky TV?
      ii is a boy?

3 Tammy did a survey of hair of all the pupils in her class. She labelled it as either long or short. This table shows her results:

	Number of boys	Number of girls
Long hair	4	11
Short hair	13	2

   a How many pupils are in Tammy's class?
   b How many girls in Tammy's class had long hair?
   c How many pupils in Tammy's class had short hair?
   d What is the probability of selecting a pupil at random that:
      i is a girl with long hair?
      ii is a short-haired pupil?

CHAPTER 16 Number 5

Practice

16A Adding and subtracting decimals

1 Without using a calculator, work out each of these.

    a 2.5 + 4.3   **b** 7.3 + 6.8
    c 21.44 + 4.79   **d** 9.3 – 6.1
    e 17.5 – 11.8   **f** 22.46 – 15.39

2 Without using a calculator, work out each of these.

    a 4.0 + 5.36   **b** 6.35 + 2.41 + 9.32
    c 0.05 + 0.9 + 1.32 + 0.83   **d** 15.81 + 7 + 19.2
    e 128.4 + 17 + 93.71

3 Without using a calculator, work out each of these.

    a 5 – 2.7   **b** 12 – 0.43   **c** 7 – 2.66   **d** 23 – 16.08

4 a How much is left if 370 mm of wire is cut from a 3 m reel? Work in metres.

    b How much is left if 187 cl of juice is poured from a jug containing 5 litres? Work in litres.

5 Lars wants to make these statues using plaster of Paris. The amount of plaster needed for each statue is shown.

bear   rhino   pig
2500 g   1.4 kg   1800 g

dog   cat
3.25 kg   2300 g

a What is the total plaster of Paris needed to make all the statues?

    Lars has a 7 kg bag of plaster of Paris. He wants to have as little plaster left over as possible.

    b Which statues should he make?
    c How much plaster would he use?
    d How much plaster will he have left over **i** in kg **ii** in g?

Practice
16B Multiplying and dividing decimals

1 Work out the following, without using a calculator.

    a 5×0.37 **b** 2.84×7 **c** 9×1.06 **d** 7.96×8

2 Work out the following, without using a calculator.

    a $6.42 \div 3$ **b** $46.06 \div 7$ **c** $23.22 \div 6$ **d** $76.59 \div 9$

 3 Calculate the total cost of 7 burgers costing £2.63 each.

 4 A motorcycle travelled 164.7 miles on 9 litres of petrol.

    a How many miles could it travel on 1 litre?
    b How many miles could it travel on 4 litres?

 5 Every second, a computer downloads 3.49 kilobytes of information from the internet. How much information does it download in 8 seconds?

 6 A pack of six mini Easter eggs weighs 152.1 g. How much does one egg weigh?

Practice
16C Using a calculator

1 Use the memory keys on your calculator to work out each of the following. Write down any values that you store in the memory.

    a $\dfrac{15.8 + 17.1}{19.2 - 9.8}$ **b** $\dfrac{92.51 - 26.75}{32 \times 0.15}$

    c $18.93 - (51.14 - 36.82)$ **d** $1.42 + \dfrac{1.5}{0.54 + 1.86}$

2 Calculate the answers to question 1 again, this time using brackets keys. Write out the key presses for each calculation. Which method uses fewer key presses?

3 Work out the value of each of these using the sign change key to enter the first negative number.

    a $-5 + 7 - 8$ **b** $-2.4 + 1.6$
    c $-17 + 7 + 17 - 7$ **d** $-132 - 46 - 105$

4 Use the square root key to work out the following.

  a √1225   b √60   c √0.6   d √0

5 a Do you need to press [×] when multiplying a bracket on your calculator?

  Try [5] [(] [3] [+] [4] [)] [=]

  b Can you leave off the last bracket of a calculation?

  Try [5] [×] [(] [3] [+] [4] [=]

Practice

16D Fractions of quantities

1 Calculate each of the following:

  a $\frac{2}{5}$ of £20   b $\frac{4}{7}$ of 56 kg   c $\frac{9}{10}$ of 110 ml
  d $1\frac{3}{4}$ of 16 eggs   e $\frac{5}{8}$ of 400 g   f $2\frac{2}{3}$ of 60 miles
  g $\frac{4}{11}$ of 121 seconds   h $\frac{19}{20}$ of 1000 tonnes

2 Calculate the following. Cancel your answers and write them as mixed fractions.

  a $4 \times \frac{3}{7}$ b $3 \times 1\frac{2}{5}$ c $7 \times \frac{3}{10}$ d $9 \times 2\frac{1}{6}$
  e $\frac{7}{8} \times 6$ f $2 \times \frac{13}{14}$ g $5 \times 3\frac{7}{10}$ h $\frac{9}{16} \times 18$

3 $\frac{4}{5}$ of the annual rainfall in Quondu occurs in June. If the annual rainfall is 260 mm, what is the rainfall in June?

4 $\frac{2}{7}$ of the questions on a maths paper are arithmetic. There are 42 questions altogether. How many are arithmetic?

FM 5 $1\frac{5}{9}$ cans of paint are needed to paint a fence. How many cans are needed to paint 6 identical fences?

Practice

16E Percentages of quantities

1 Calculate the following:

  a 80% of £2.40   b 65% of 80 kg   c 5% of £3200
  d 44% of 600 people

2 Copy these calculations.

  a … % of 40 = 18
  b … % of 300 = 96
  c … % of 250 = 100

The missing percentages are 32%, 40% and 45%.
Complete your calculations with the correct percentages.

3 A bag contains 200 chocolate beans. Anna eats 20% and then gives the bag to Raj. Then, Raj eats 35% of the remaining beans. Siobhan gets what is left over.

    a How many beans does Anna eat?
    b How many beans does she give to Raj?
    c How many beans does Raj eat?
    d How many beans does Siobhan get?
    e What percentage of the original bag does Siobhan get?

4 Copy and complete this table.

	a	b	c	d	e
Decimal	0.8			0.875	
Fraction			$\frac{7}{20}$	$\frac{19}{25}$	
Percentage		6%			

CHAPTER 17 Algebra 5

Practice 17A Solving equations

1 Solve the following equations.

    a $2y + 1 = 11$ **b** $3b + 2 = 14$ **c** $7x + 6 = 41$
    d $10p + 40 = 100$ **e** $6q + 8 = 56$ **f** $2m + 100 = 400$
    g $5d + 17 = 52$ **h** $8r + 28 = 92$

2 Solve the following equations.

    a $4m - 3 = 13$ **b** $2q - 7 = 11$ **c** $5v - 10 = 25$
    d $7n - 13 = 8$ **e** $10A - 60 = 100$ **f** $2w - 35 = 35$
    g $6f - 18 = 36$ **h** $11x - 17 = 49$

3 Solve each of the following equations.

    a $3w - 8 = 25$ **b** $5e - 30 = 10$ **c** $12m - 4 = 92$
    d $6v + 4 = 28$ **e** $9j - 17 = 55$ **f** $30x + 100 = 700$

4 Two of these equations have the same solution. Find them. Write down the solutions to all of the equations.

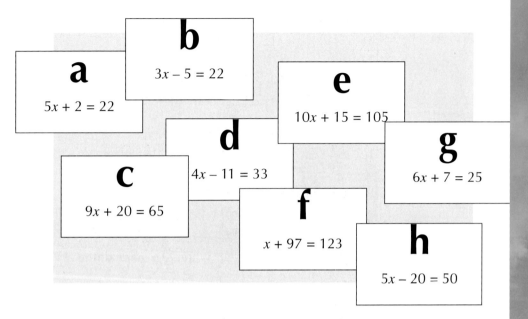

a $5x + 2 = 22$
b $3x - 5 = 22$
c $9x + 20 = 65$
d $4x - 11 = 33$
e $10x + 15 = 105$
f $x + 97 = 123$
g $6x + 7 = 25$
h $5x - 20 = 50$

5 Solve each of the following equations using number machines.

Example

$5x + 2 = 37$

Working Write the equation using number machines:

$x \rightarrow \boxed{\times 5} \rightarrow \boxed{+ 2} \rightarrow 37$

$x \leftarrow \boxed{\div 5} \leftarrow \boxed{- 2} \leftarrow 37$ (reverse the number machines)

$x \leftarrow \boxed{\div 5} \leftarrow 35$ (work backwards to find the answer)

$x = 7$

a $4p + 11 = 47$ **b** $6v - 23 = 97$ **c** $10n + 100 = 1000$

Practice

17B Formulae

1 This formula is used to generate odd numbers: $m = 2n + 1$. Find m when:

a $n = 4$ **b** $n = 16$

2 This formula can be used to convert pounds (£) into Euros (€):

$E = \dfrac{8P}{5}$

where P is the number of pounds and E is the number of Euros.

a Use the formula to convert £40 to Euros.
b Use the formula to convert £18 to Euros.

3 The time needed to roast a joint of meat is given by the formula:

$T = 30w + E$

where T is the cooking time (in minutes), w is the weight of the joint, in kg, and E is extra time.

   a Calculate the cooking time for a joint weighing 2.5 kg which needs 25 minutes of extra time.
   b Calculate the cooking time when $w = 1.8$ kg and $E = 15$.

4 This metal plate was made by cutting a square hole from a square piece of metal. Its perimeter is given by the formula:

$P = 4(D + d)$

where P is the perimeter, D is the length of the outer side and d is the length of the inner side.

   a Calculate the perimeter when the outer side is 25 mm and the inner side is 18 mm.
   b Calculate the perimeter when $D = 4.6$ cm and $d = 1.25$ cm.

Practice 17C A square investigation

Each shape is made using an inner square and an outer square. The side of the outer square is shown below each shape.

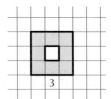 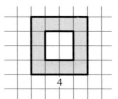

1 Copy this table.

Side of outer square, s	3	4	5	6	7	8
Number of shaded squares, n						

2 Count the numbers of shaded squares and write them in your table.

3 Draw more shapes to complete the table. The inner square must be as large as possible.

4 What do you notice about the numbers of shaded squares in the table?

5 Write down a rule that gives the number of shaded squares, n, if you are told the side, s. Hint: Start by multiplying by 4.

6 Write your rule using algebra.

Practice 17D Graphs from the real world

  1 The graph shows how long it takes for a computer printer to print a number of photos. Use the conversion graph to answer the following questions.

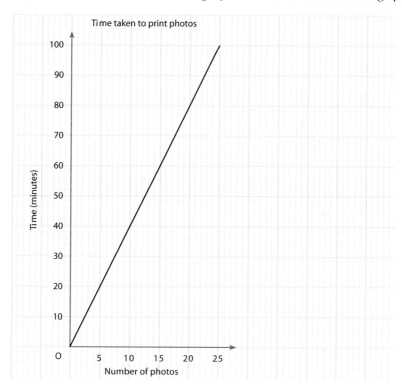

a How long does it take to print i 10 photos ii 22 photos?
b How many photos can be printed in i 80 minutes ii 48 minutes?

  2 A clothing mail order company charges £2.50 per item for postage and packing.

a Copy and complete the table.

Number of items	2	4	6	8	10	12	14
Postage & packing (£)			15				

b Draw a graph using these scales:
  x-axis (Number of items)   2 cm to 2 items
  y-axis (Postage & packing) 2 cm to £5
  Use your graph to answer the following questions.
c What is the postage and packing charge for i 6 items ii 13 items?
d How many items can be delivered for a postage and packing charge of
  i £10 ii £27.50?

  3 The hard cover of a hardback novel weighs 50 g. 20 pages weigh 40 g.

a Copy and complete the table.

Number of pages	100	200	300	400
Weight of book (g)			650	

b Draw a graph using these scales:
      x-axis (Number of pages)   2 cm to 100 pages
      y-axis (Weight of book)   2 cm to 100 g
      Use your graph to answer the following questions.
   c What is the weight of a novel with   i 150 pages   ii 320 pages?
   d How many pages does a novel have that weighs   i 730 g   ii 590 g?

Geometry and Measures 5

18A Polygons

1
 a   b   c   d   e

   i   Name each polygon, for example, heptagon.
   ii   Which polygons are regular?
   iii Describe each polygon as concave or convex.

2 Draw these polygons.

   a A pentagon with one right-angle and one reflex angle.
   b A hexagon with rotational symmetry, order 2.
   c An octagon with exactly two lines of symmetry.
   d A hexagon with one line of symmetry and two reflex angles.

3 Use squared paper to draw 8 different kinds of polygon inside a 3 by 3 grid. Use a different grid for each polygon. Describe each polygon, for example, convex quadrilateral.

18B Tessellations

Make a tessellation from each of the following shapes, if possible. Use a square grid to help.

Practice 18C Constructing 3-D shapes

1. Which of the following are nets for the cuboid shown?

   a 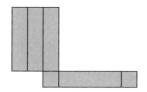 b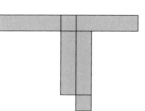

   c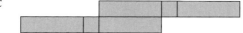

2. Which of the following are nets for the half cylinder shown?

   a   b   c   d

3. Stick a photocopy or a tracing of this net on to a sheet of thin card. Cut out around the net shape. Fold and glue it to make the box.

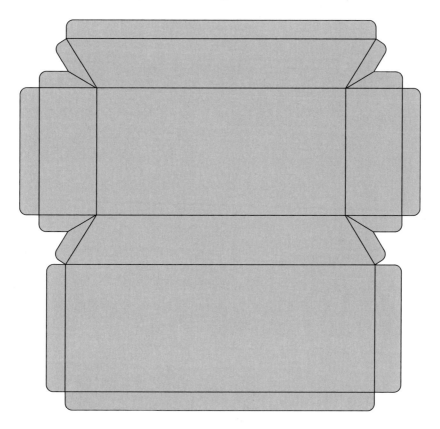